JACOB RAPOO TLHAGALE

SWIM OR SINK-LIFE: THE SURVIVAL OF THE FITTEST

JACOB RAPOO TLHAGALE

SWIM OR SINK-LIFE: THE SURVIVAL OF THE FITTEST

LIFE IS FULL OF UPS AND DOWNS AND LOTS OF CHALLENGES - TO BE ABLE TO SURVIVE ITS EITHER YOUR SWIM OR YOU SINK

Blessed Hope Publishing

Cover image: www.ingimage.com

Publisher:
Blessed Hope Publishing
is a trademark of
International Book Market Service Ltd., member of OmniScriptum Publishing Group
17 Meldrum Street, Beau Bassin 71504, Mauritius
Printed at: see last page
ISBN: 978-613-7-92466-2

TITTLE : "SWIM OR SINK:LIFE =SURVIVAL OF THE FITTEST.'

Survival of the fittest" is a phrase that originated from Darwinian evolutionary theory as a way of describing the mechanism of natural selection. The biological concept of fitness is defined as reproductive success. In Darwinian terms the phrase is best understood as "Survival of the form that will leave the most copies of itself in successive generations."

Herbert Spencer first used the phrase, after reading Charles Darwin's *On the Origin of Species*, in his *Principles of Biology* (1864), in which he drew parallels between his own economic theories and Darwin's biological ones: "This survival of the fittest, which I have here sought to express in mechanical terms, is that which Mr. Darwin has called 'natural selection', or the preservation of favoured races in the struggle for life."

Darwin responded positively to Alfred Russel Wallace's suggestion of using Spencer's new phrase "survival of the fittest" as an alternative to "natural selection", and adopted the phrase in *The Variation of Animals and Plants under Domestication* published in 1868. In *On the Origin of Species*, he introduced the phrase in the fifth edition published in 1869, intending it to mean "better designed for an immediate, local environment".

By his own account, Herbert Spencer described a concept similar to "survival of the fittest" in his 1852 "A Theory of Population".[7] He first used the phrase – after reading Charles Darwin's *On the Origin of Species* – in his *Principles of Biology* of 1864 in which he drew parallels between his economic theories and Darwin's biological, evolutionary ones, writing, "This survival of the fittest, which I have here sought to express in mechanical terms, is that which Mr. Darwin has called 'natural selection', or the preservation of favored races in the struggle for life."

In July 1866 Alfred Russel Wallace wrote to Darwin about readers thinking that the phrase "natural selection" personified nature as "selecting", and said this misconception could be avoided "by adopting Spencer's term" *Survival of the fittest*. Darwin promptly replied that Wallace's letter was "as clear as daylight. I fully agree with all that you say on the advantages of H. Spencer's excellent expression of 'the survival of the fittest'. This however had not occurred to me till reading your letter. It is, however, a great objection to this term that it cannot be used as a substantive governing a verb". Had he received the letter two months earlier, he would have worked the phrase into the fourth edition of the *Origin* which was then being printed, and he would use it in his "next book on Domestic Animals etc.".

Darwin wrote on page 6 of *The Variation of Animals and Plants under Domestication* published in 1868, "This preservation, during the battle for life, of varieties which possess any advantage in structure, constitution, or instinct, I have called Natural Selection; and Mr. Herbert Spencer has well expressed the same idea by the Survival of the Fittest. The term "natural selection" is in some respects a bad one, as it seems to imply conscious choice; but this will be disregarded after a little familiarity". He defended his analogy as similar to language used in chemistry, and to astronomers depicting the "attraction of gravity as ruling the movements of the planets", or the way in which "agriculturists speak of man making domestic races by his power of selection". He had "often personified the word Nature; for I have found it difficult to avoid this ambiguity; but I mean by nature only the aggregate action and product of many natural laws,—and by laws only the ascertained sequence of events."

In the first four editions of *On the Origin of Species*, Darwin had used the phrase "natural selection". In Chapter 4 of the 5th edition of *The Origin* published in 1869, Darwin implies again the synonym: "Natural Selection, or the Survival of the Fittest". By "fittest" Darwin meant "better adapted for the immediate, local environment", not the common modern meaning of "in the best physical shape" (think of a puzzle piece, not an athlete).[5] In the introduction he gave full credit to Spencer, writing "I have called this principle, by which each slight variation, if useful, is preserved, by the term Natural Selection, in order to mark its relation to man's power of selection. But the expression often used by Mr. Herbert Spencer of the Survival of the Fittest is more accurate, and is sometimes equally convenient.

In *The Man Versus The State*, Spencer used the phrase in a postscript to justify a plausible explanation of how his theories would not be adopted by "societies of militant type". He uses the term in the context of societies at war, and the form of his reference suggests that he is applying a general principle.

"Thus by survival of the fittest, the militant type of society becomes characterized by profound confidence in the governing power, joined with a loyalty causing submission to it in all matters whatever".

Though Spencer's conception of organic evolution is commonly interpreted as a form of Lamarckism, Herbert Spencer is sometimes credited with inaugurating Social Darwinism. The phrase "survival of the fittest" has become widely used in popular literature as a catchphrase for any topic related or analogous to evolution and natural selection. It has thus been applied to principles of unrestrained competition, and it has been used extensively by both proponents and opponents of Social Darwinism.]

Evolutionary biologists criticise the manner in which the term is used by non-scientists and the connotations that have grown around the term in popular culture. The phrase also does not help in conveying the complex nature of natural selection, so modern biologists prefer and almost exclusively use the term natural selection. The biological concept of fitness refers to reproductive success, as opposed to survival, and is not explicit in the specific ways in which organisms can be more "fit" (increase reproductive success) as having phenotypic characteristics that enhance survival and reproduction (which was the meaning that Spencer had in mind).

While the phrase "survival of the fittest" is often used to mean "natural selection", it is avoided by modern biologists, because the phrase can be misleading. For example, survival is only one aspect of selection, and not always the most important. Another problem is that the word "fit" is frequently confused with a state of physical fitness. In the evolutionary meaning "fitness" is the rate of reproductive output among a class of genetic variants.

The phrase can also be interpreted to express a theory or hypothesis: that "fit" as opposed to "unfit" individuals or species, in some sense of "fit", will survive some test. Nevertheless, when extended to individuals it is a conceptual mistake, the phrase is a reference to the *transgenerational* survival of the heritable attributes; particular individuals are quite irrelevant. This becomes more clear when referring to Viral quasispecies, in *survival of the flattest*, which makes it clear *to survive* makes no reference to the question of even being alive itself; rather the functional capacity of proteins to carry out work.

Interpretations of the phrase as expressing a theory are in danger of being tautological, meaning roughly "those with a propensity to survive have a propensity to survive"; to have content the theory must use a concept of fitness that is independent of that of survival.

Interpreted as a theory of species survival, the theory that the fittest species survive is undermined by evidence that while direct competition is observed between individuals, populations and species, there is little evidence that competition has been the driving force in the evolution of large groups such as, for example, amphibians, reptiles, and mammals. Instead, these groups have evolved by expanding into empty ecological niches.[16] In the punctuated equilibrium model of environmental and biological change, the factor determining survival is often not superiority over another in competition but ability to survive dramatic changes in environmental conditions, such as after a meteor impact energetic enough to greatly change the environment globally. The main land dwelling animals to survive the K-Pg impact 66 million years ago had the ability to live in tunnels, for example.

In 2010 Sahney et al. argued that there is little evidence that intrinsic, biological factors such as competition have been the driving force in the evolution of large groups. Instead, they cited extrinsic, abiotic factors such as expansion as the driving factor on a large evolutionary scale. The rise of dominant groups such as amphibians, reptiles, mammals and birds occurred by opportunistic expansion into empty ecological niches and the extinction of groups happened due to large shifts in the abiotic environment.

Interpreted as expressing a moral theory]

SOCIAL DARWINISTS

It has been claimed that "the survival of the fittest" theory in biology was interpreted by late 19th century capitalists as "an ethical precept that sanctioned cut-throat economic competition" and led to the advent of the theory of "social Darwinism" which was used to justify laissez-faire economics, war and racism. However, these ideas predate and commonly contradict Darwin's ideas, and indeed their proponents rarely invoked Darwin in support.[citation needed] The term "social Darwinism" referring to capitalist ideologies was introduced as a term of abuse by Richard Hofstadter's *Social Darwinism in American Thought* published in 1944.

ANARCHISTS

Russian anarchist Peter Kropotkin viewed the concept of "survival of the fittest" as supporting co-operation rather than competition. In his book *Mutual Aid: A Factor of Evolution* he set out his analysis leading to the conclusion that the fittest was not necessarily the best at competing individually, but often the community made up of those best at working together. He concluded that

In the animal world we have seen that the vast majority of species live in societies, and that they find in association the best arms for the struggle for life: understood in its wide Darwinian sense – not as a struggle for the sheer means of existence, but as a struggle against all natural conditions unfavorable to the species. The animal species, in which individual struggle has been reduced to its narrowest limits, and the practice of mutual aid has attained the greatest development, are invariably the most numerous, the most prosperous, and the most open to further progress.

Applying this concept to human society, Kropotkin presented mutual aid as one of the dominant factors of evolution, the other being self-assertion, and concluded that

In the practice of mutual aid, which we can retrace to the earliest beginnings of evolution, we thus find the positive and undoubted origin of our ethical conceptions; and we can affirm that in the ethical progress of man, mutual support not mutual struggle – has had the leading part. In its wide extension, even at the present time, we also see the best guarantee of a still loftier evolution of our race.

TAUTOLOGY

"Survival of the fittest" is sometimes claimed to be a tautology. The reasoning is that if one takes the term "fit" to mean "endowed with phenotypic characteristics which improve chances of survival and reproduction" (which is roughly how Spencer understood it), then "survival of the fittest" can simply be rewritten as "survival of those who are better equipped for surviving". Furthermore, the expression *does* become a tautology if one uses the most widely accepted definition of "fitness" in modern biology, namely reproductive success itself (rather than any set of characters conducive to this reproductive success). This reasoning is sometimes used to claim that Darwin's entire theory of evolution by natural selection is fundamentally tautological, and therefore devoid of any explanatory power.

However, the expression "survival of the fittest" (taken on its own and out of context) gives a very incomplete account of the mechanism of natural selection. The reason is that it does not mention a key requirement for natural selection, namely the requirement of *heritability*. It is true that the phrase "survival of the fittest", in and by itself, is a tautology if fitness is defined by survival and reproduction. Natural selection is the portion of variation in reproductive success that is caused by *heritable* characters (see the article on natural selection).

If certain heritable characters increase or decrease the chances of survival and reproduction of their bearers, then it follows mechanically (by definition of "heritable") that those characters that improve survival and reproduction will increase in frequency over generations. This is precisely what is called "evolution by natural selection". On the other hand, if the characters which lead to differential reproductive success are not heritable, then no meaningful evolution will occur, "survival of the fittest" or not: if improvement in reproductive success is caused by traits that are not heritable, then there is no reason why these traits should increase in frequency over generations. In other words, natural selection does not simply state that "survivors survive" or "reproducers reproduce"; rather, it states that "survivors survive, reproduce and *therefore* propagate any *heritable* characters which have affected their survival and reproductive success". This statement is not tautological: it hinges on the testable hypothesis that such fitness-impacting heritable variations actually exist (a hypothesis that has been amply confirmed.)

Momme von Sydow suggested further definitions of 'survival of the fittest' that may yield a testable meaning in biology and also in other areas where Darwinian processes have been influential. However, much care would be needed to disentangle tautological from testable aspects. Moreover, an "implicit shifting between a testable and an untestable interpretation can be an illicit tactic to immunize natural selection ... while conveying the impression that one is concerned with testable hypotheses".

Skeptic Society founder and *Skeptic* magazine publisher Michael Shermer addresses the tautology problem in his 1997 book, *Why People Believe Weird Things*, in which he points out that although tautologies are sometimes the beginning of science, they are never the end, and that scientific principles like natural selection are testable and falsifiable by virtue of their predictive power. Shermer points out, as an example, that population genetics accurately demonstrate when natural selection will and will not effect change on a population. Shermer hypothesizes that if hominid fossils were found in the same geological strata as trilobites, it would be evidence against natural selection.

SURVIVAL OF THE FITTEST QUOTES

“It is not the strongest or the most intelligent who will survive but those who can best manage change.”
– **Leon C. Megginson**

“The weak die out and the strong will survive, and will live on forever”
– **Anne Frank, The Diary of Anne Frank**

“The law of evolution is that the strongest survives!' 'Yes, and the strongest, in the existence of any social species, are those who are most social. In human terms, most ethical...There is no strength to be gained from hurting one another. Only weakness.”
– **Ursula K. Le Guin**

“Survivors aren't always the strongest; sometimes they're the smartest, but more often simply the luckiest.”
– **Carrie Ryan, The Dark and Hollow Places**

“He was a killer, a thing that preyed, living on the things that lived, unaided, alone, by virtue of his own strength and prowess, surviving triumphantly in a hostile environment where only the strong survive.”
– **Jack London, The Call of the Wild**

“Institutionalized rejection of difference is an absolute necessity in a profit economy which needs outsiders as surplus people.”
– **Audre Lorde**

“With savages, the weak in body or mind are soon eliminated; and those that survive commonly exhibit a vigorous state of health. We civilised men, on the other hand, do our utmost to check the process of elimination; we build asylums for the imbecile, the maimed, and the sick; we institute poor-laws; and our medical men exert their utmost skill to save the life of every one to the last moment. There is reason to believe that vaccination has preserved thousands, who from a weak constitution would formerly have succumbed to small-pox. Thus the weak members of civilized societies propagate their kind. No one who has attended to the breeding of domestic animals will doubt that this must be highly injurious to the race of man. It is surprising how soon a want of care, or care wrongly directed, leads to the degeneration of a domestic race; but excepting in the case of man himself, hardly any one is so ignorant as to allow his worst animals to breed.

The aid which we feel impelled to give to the helpless is mainly an

incidental result of the instinct of sympathy, which was originally acquired as part of the social instincts, but subsequently rendered, in the manner previously indicated, more tender and more widely diffused. Nor could we check our sympathy, if so urged by hard reason, without deterioration in the noblest part of our nature. The surgeon may harden himself whilst performing an operation, for he knows that he is acting for the good of his patient; but if we were intentionally to neglect the weak and helpless, it could only be for a contingent benefit, with a certain and great present evil. Hence we must bear without complaining the undoubtedly bad effects of the weak surviving and propagating their kind; but there appears to be at least one check in steady action, namely the weaker and inferior members of society not marrying so freely as the sound; and this check might be indefinitely increased, though this is more to be hoped for than expected, by the weak in body or mind refraining from marriage."
– **Charles Darwin**

"This century will be called Darwin's century. He was one of the greatest men who ever touched this globe. He has explained more of the phenomena of life than all of the religious teachers. Write the name of Charles Darwin on the one hand and the name of every theologian who ever lived on the other, and from that name has come more light to the world than from all of those. His doctrine of evolution, his doctrine of the survival of the fittest, his doctrine of the origin of species, has removed in every thinking mind the last vestige of orthodox Christianity. He has not only stated, but he has demonstrated, that the inspired writer knew nothing of this world, nothing of the origin of man, nothing of geology, nothing of astronomy, nothing of nature; that the Bible is a book written by ignorance--at the instigation of fear. Think of the men who replied to him. Only a few years ago there was no person too ignorant to successfully answer Charles Darwin, and the more ignorant he was the more cheerfully he undertook the task. He was held up to the ridicule, the scorn and contempt of the Christian world, and yet when he died, England was proud to put his dust with that of her noblest and her grandest. Charles Darwin conquered the intellectual world, and his doctrines are now accepted facts. His light has broken in on some of the clergy, and the greatest man who to-day occupies the pulpit of one of the orthodox churches, Henry Ward Beecher, is a believer in the theories of Charles Darwin--a man of more genius than all the clergy of that entire church put together.

...The church teaches that man was created perfect, and that for six thousand years he has degenerated. Darwin demonstrated the falsity of this dogma. He shows that man has for thousands of ages steadily advanced; that the Garden of Eden is an ignorant myth; that the doctrine of original sin has no foundation in fact; that the atonement is an absurdity; that the serpent did not tempt, and that man did not 'fall.'

Charles Darwin destroyed the foundation of orthodox Christianity. There is nothing left but faith in what we know could not and did not happen. Religion and science are enemies. One is a superstition; the other is a fact. One rests upon the false, the other upon the true. One is the result of fear and faith, the other of investigation and reason."— **Robert Green Ingersoll**

"The very comprehensibility of the world points to an intelligence behind the world. Indeed, science would be impossible if our intelligence were not adapted to the intelligibility of the world. The match between our intelligence and the intelligibility of the world is no accident. Nor can it properly be attributed to natural selection, which places a premium on survival and reproduction and has no stake in truth or conscious thought. Indeed, meat-puppet robots are just fine as the output of a Darwinian evolutionary process."— **William A. Dembski**

"There was no such thing as a fair fight. All vulnerabilities must be exploited."— **Cary Caffrey**

"The spirit finds a way to be born.Instinct seeks for ways to urvive."— **Toba Beta,**

"If theft is advantageous to everyone who succeeds at it, and adultery is a good strategy, at least for males, for increasing presence in the gene pool, why do we feel they are wrong? Shouldn't the only morality that evolution produces be the kind Bill Clinton had - being sorry you got caught?"
— **Robert J. Sawyer**

"My firm conviction is that if wide-spread Eugenic reforms are not adopted during the next hundred years or so, our Western Civilization is inevitably destined to such a slow and gradual decay as that which has been experienced in the past by every great ancient civilization. The size and the importance of the United States throws on you a special responsibility in your endeavours to safeguard the future of our race. Those who are attending your Congress will be aiding in this endeavour, and though you will gain no thanks from your own generation, posterity will, I believe, learn to realize the great dept it owes to all the workers in this field."– **Leonard Darwin**

"We are all in the middle of nature beauty contest."– **Toba Beta**

"If anyone rises to power, it's not only because he could, but also because the stars were aligned in his favor. Many with apparent means to take it failed simply because they weren't destined for the honor"
– **Bangambiki Habyarimana, Pearls Of Eternity**

"When one faces pain on a daily basis, one either learns to live with it or let it consume him."– **Jina S. Bazzar**

"There were sharks before there were dinosaurs, and the reason sharks are still in the ocean is that nothing is better at being a shark than a shark."
– **Eliot Peper, Bandwidth**

"Weedon Scott had set himself the task of redeeming White Fang - or rather, of redeeming mankind from the wrong it had done White Fang. It was a matter of principle and conscience. He felt that the ill done White Fang was a debt incurred by man and that it must be paid."– **Jack London**

"Why should he not hate them? He never asked himself the question. He knew only hate and lost himself in the passion of it. Life had become a hell to him. He had not been made for the close confinement wild beasts endure at the hands of men. And yet it was in precisely this way that he was treated. Men stared at him, poked sticks between the bars to make him snarl, and then laughed at him."– **Jack London**

"but oftener he remembered the man in the red sweater, the death of Curly, the great fight with Spitz and the good things he had eaten or would like to eat. He was not homesick. The Sunland was very dim and distant, and such memories had no power over him. Far more potent were the memories of his heredity that gave things he had never seen before a seeming familiarity; the instincts (which were but the memories of his ancestors become habits) which had lapsed in later days, and still later, in him, quickened and became alive again."– **Jack London**

"Civilization is much more than the survival of the fittest and the unrelenting culling of the weakest members. Civilized people share a value system that extends far beyond doing whatever it takes to survive. Mere barbarians might be devoted to a life of exploitation. In contrast, civilized people value nature and care for the most vulnerable members of their kind."– **Kilroy J. Oldster**

"Weak prey didn't live long. It was the law of the jungle."– **Jina S. Bazzar**

"The survival of the fittest instinct should end with our generation. We have evolved to grow out of it. There are enough resources out there for us not to kill someone over them."– **Rajesh**

"...As Joe walked back to them, he heard the cow elk gurgle and settle into the grass on top of her calf.

"This is what wolves do," Joe said, his voice calm, a betrayal of what he felt. "I'm not saying they shouldn't be here, but this is what they do. *They're wolves*. I know it sounds real nice to say they're magical and beautiful and they balance nature and restore an ecosystem--and it's true, they do that. But this is how they do it. They go after the weakest first. When the mother stays back, the wolves open a hole in her belly and pull out her entrails. Then they wait until she doesn't have the strength to protect herself, then they'll move in and tear her throat out."– **C.J. Box**

"The biologists are a ruthless bunch, and quite devoid of any particular regard for doctors. In theory they don't even want a surgeon to patch up a man who has been hit on the head with a brick, the argument being that it is better for the human race to be able to dodge bricks. Which is a correct viewpoint, I suppose, unless it happens to be your head and your brick."
– **Dr. Blake F. Donaldson, Strong Medicine**

"Tig would have fit on that plane just fine. Tig, Art, Takis, these anomalous, scrappy survivors, might be the lucky ones. They ate less and took up less space: the humans of the future."– **Barbara Kingsolver**

"Survival of the fittest," which Sax had always considered a useless tautology. But if social Darwinists were taking over, then maybe the concept gained importance, as a religious dogma of the ruling order...."– **Kim Stanley Robinson,**

"After the lockdown, when markets become less active, the subject of mainstream economics faces another tough ground. There are millions of poor people who do not have work. When lockdown happens, a great many people find resource markets stalled where they used to get income. More than ever, such crises necessitate the flow of resources from the haves to the have-nots. But, frozen goods and resource markets cannot help much, especially the poor and vulnerable people. That is where, pro-social behaviour and beyond-market distribution of resources is necessary. However, mainstream economics treats altruism as 'impure'. It looks at altruism in the paradigm of pursuing self-interest."— **Salman Ahmed Shaikh**

"The paradigm of self-interest is neutral between a person's decision to help others or to not help others. If fear and uncertainty make people more short-sighted, self-centric and engage in hoarding cash, withdrawing from banks, disinvesting their savings in capital markets and buying essential goods in bulk, then there is no drive, urge or impetus that economics can offer to suggest otherwise. It is neutral between these choices and the choice for altruism.

On the other hand, as per physics, we were stardust before and stardust after extinction. As per evolutionary biology, we are one of the millions and millions of species that have earned the right to survive under the sun for a brief period until genetic mutations bury our specie as well. We were born through a fierce and destructive competition and survive until we manage to withstand that competition. While waiting and acting on morals, we should not be here in the first place."— **Salman Ahmed Shaikh**

"Motivation for moral behaviour and pro-social cooperative response has to come from somewhere else. In poor countries like Pakistan, people with surplus resources are engaging in charity, donations, and volunteering. Empirical evidence in Pakistan in multiple research studies has found that faith is the biggest motivation behind charitable donations and it encapsulates and is associated with other humane motives. This trend is also seen in other parts of the world. But, economics is largely silent and irrelevant when it comes to exchange, allocation and distribution of economic resources outside of markets."— **Salman Ahmed Shaikh,**

STEADFASTNESS IN THE BIBLE

1 Corinthians 15:58 Therefore, my beloved brothers, be steadfast, immovable, always abounding in the work of the Lord, knowing that in the Lord your labor is not in vain.

James 1:12 Blessed is the man who remains steadfast under trial, for when he has stood the test he will receive the crown of life, which God has promised to those who love him.

James 1:2-4 Count it all joy, my brothers, when you meet trials of various kinds, for you know that the testing of your faith produces steadfastness. And let steadfastness have its full effect, that you may be perfect and complete, lacking in nothing.

Psalm 112:7 He is not afraid of bad news; his heart is firm, trusting in the Lord.

2 Peter 1:5-8 For this very reason, make every effort to supplement your faith with virtue, and virtue with knowledge, and knowledge with self-control, and self-control with steadfastness, and steadfastness with godliness, and godliness with brotherly affection, and brotherly affection with love. For if these qualities are yours and are increasing, they keep you from being ineffective or unfruitful in the knowledge of our Lord Jesus Christ.

Titus 2:2 Older men are to be sober-minded, dignified, self-controlled, sound in faith, in love, and in steadfastness.

Isaiah 40:31 But they who wait for the Lord shall renew their strength; they shall mount up with wings like eagles; they shall run and not be weary; they shall walk and not faint.

Colossians 1:1-29 Paul, an apostle of Christ Jesus by the will of God, and Timothy our brother, To the saints and faithful brothers in Christ at Colossae: Grace to you and peace from God our Father. We always thank God, the Father of our Lord Jesus Christ, when we pray for you, since we heard of your faith in Christ Jesus and of the love that you have for all the saints, because of the hope laid up for you in heaven. Of this you have heard before in the word of the truth, the gospel, ...

2 Corinthians 3:6 Who has made us competent to be ministers of a new covenant, not of the letter but of the Spirit. For the letter kills, but the Spirit gives life.

Hebrews 6:18 So that by two unchangeable things, in which it is impossible for God to lie, we who have fled for refuge might have strong encouragement to hold fast to the hope set before us.

John 3:16 "For God so loved the world, that he gave his only Son, that whoever believes in him should not perish but have eternal life.

Romans 8:18-25 For I consider that the sufferings of this present time are not worth comparing with the glory that is to be revealed to us. For the creation waits with eager longing for the revealing of the sons of God. For the creation was subjected to futility, not willingly, but because of him who subjected it, in hope that the creation itself will be set free from its bondage to corruption and obtain the freedom of the glory of the children of God. For we know that the whole creation has been groaning together in the pains of childbirth until now. ...

Romans 8:2 For the law of the Spirit of life has set you free in Christ Jesus from the law of sin and death.

Galatians 6:9 And let us not grow weary of doing good, for in due season we will reap, if we do not give up.

Isaiah 26:3 You keep him in perfect peace whose mind is stayed on you, because he trusts in you.

Deuteronomy 7:9 Know therefore that the Lord your God is God, the faithful God who keeps covenant and steadfast love with those who love him and keep his commandments, to a thousand generations,

Psalm 51:10 Create in me a clean heart, O God, and renew a right spirit within me.

Galatians 2:20

I have been crucified with Christ. It is no longer I who live, but Christ who lives in me. And the life I now live in the flesh I live by faith in the Son of God, who loved me and gave himself for me.

Lamentations 3:22-23 The steadfast love of the Lord never ceases; his mercies never come to an end; they are new every morning; great is your faithfulness.

Psalm 86:15 But you, O Lord, are a God merciful and gracious, slow to anger and abounding in steadfast love and faithfulness.

Colossians 1:23 If indeed you continue in the faith, stable and steadfast, not shifting from the hope of the gospel that you heard, which has been proclaimed in all creation under heaven, and of which I, Paul, became a minister.

Romans 5:8 But God shows his love for us in that while we were still sinners, Christ died for us.

1 John 3:1See what kind of love the Father has given to us, that we should be called children of God; and so we are. The reason why the world does not know us is that it did not know him.

Daniel 2:44 And in the days of those kings the God of heaven will set up a kingdom that shall never be destroyed, nor shall the kingdom be left to another people. It shall break in pieces all these kingdoms and bring them to an end, and it shall stand forever,

Joshua 1:9 Have I not commanded you? Be strong and courageous. Do not be frightened, and do not be dismayed, for the Lord your God is with you wherever you go."

1 John 4:7-8 Beloved, let us love one another, for love is from God, and whoever loves has been born of God and knows God. Anyone who does not love does not know God, because God is love.

1 Peter 5:7 Casting all your anxieties on him, because he cares for you.

Hebrews 11:1 Now faith is the assurance of things hoped for, the conviction of things not seen.

Hebrews 10:23 Let us hold fast the confession of our hope without wavering, for he who promised is faithful.

Ephesians 2:4-5 But God, being rich in mercy, because of the great love with which he loved us, even when we were dead in our trespasses, made us alive together with Christ—by grace you have been saved—

Zephaniah 3:17 The Lord your God is in your midst, a mighty one who will save; he will rejoice over you with gladness; he will quiet you by his love; he will exult over you with loud singing.

Psalm 136:1-26 Give thanks to the Lord, for he is good, for his steadfast love endures forever. Give thanks to the God of gods, for his steadfast love endures forever. Give thanks to the Lord of lords, for his steadfast love endures forever; to him who alone does great wonders, for his steadfast love endures forever; to him who by understanding made the heavens, for his steadfast love endures forever; ...

2 Thessalonians 3:5 May the Lord direct your hearts to the love of God and to the steadfastness of Christ.

Colossians 1:13 He has delivered us from the domain of darkness and transferred us to the kingdom of his beloved Son,

Philippians 4:6Do not be anxious about anything, but in everything by prayer and supplication with thanksgiving let your requests be made known to God.

Philippians 2:12 Therefore, my beloved, as you have always obeyed, so now, not only as in my presence but much more in my absence, work out your own salvation with fear and trembling,

Ephesians 4:28 Let the thief no longer steal, but rather let him labor, doing honest work with his own hands, so that he may have something to share with anyone in need.

Ephesians 4:14 So that we may no longer be children, tossed to and fro by the waves and carried about by every wind of doctrine, by human cunning, by craftiness in deceitful schemes.

Luke 1:37 For nothing will be impossible with God."

Isaiah 41:10 Fear not, for I am with you; be not dismayed, for I am your God; I will strengthen you, I will help you, I will uphold you with my righteous right hand.

2 Peter 3:17 You therefore, beloved, knowing this beforehand, take care that you are not carried away with the error of lawless people and lose your own stability.

1 Peter 5:9 Resist him, firm in your faith, knowing that the same kinds of suffering are being experienced by your brotherhood throughout the world.

Colossians 2:12 Having been buried with him in baptism, in which you were also raised with him through faith in the powerful working of God, who raised him from the dead.

Romans 8:37-39 No, in all these things we are more than conquerors through him who loved us. For I am sure that neither death nor life, nor angels nor rulers, nor things present nor things to come, nor powers, nor height nor depth, nor anything else in all creation, will be able to separate us from the love of God in Christ Jesus our Lord.

Acts 20:29 I know that after my departure fierce wolves will come in among you, not sparing the flock;

John 14:27 Peace I leave with you; my peace I give to you. Not as the world gives do I give to you. Let not your hearts be troubled, neither let them be afraid.

Matthew 28:20 Teaching them to observe all that I have commanded you. And behold, I am with you always, to the end of the age."

Psalm 136:26 Give thanks to the God of heaven, for his steadfast love endures forever.

Revelation 21:1-27 Then I saw a new heaven and a new earth, for the first heaven and the first earth had passed away, and the sea was no more. And I saw the holy city, new Jerusalem, coming down out of heaven from God, prepared as a bride adorned for her husband. And I heard a loud voice from the throne saying, "Behold, the dwelling place of God is with man. He will dwell with them, and they will be his people, and God himself will be with them as their God. He will wipe away every tear from their eyes, and death shall be no more, neither shall there be mourning, nor crying, nor pain anymore, for the former things have passed away." And he who was seated on the throne said, "Behold, I am making all things new." Also he said, "Write this down, for these words are trustworthy and true." ...

1 John 4:9-11In this the love of God was made manifest among us, that God sent his only Son into the world, so that we might live through him. In this is love, not that we have loved God but that he loved us and sent his Son to be the propitiation for our sins. Beloved, if God so loved us, we also ought to love one another.

Hebrews 6:1-20Therefore let us leave the elementary doctrine of Christ and go on to maturity, not laying again a foundation of repentance from dead works and of faith toward God, and of instruction about washings, the laying on of hands, the resurrection of the dead, and eternal judgment. And this we will do if God permits. For it is impossible, in the case of those who have once been enlightened, who have tasted the heavenly gift, and have shared in the Holy Spirit, and have tasted the goodness of the word of God and the powers of the age to come, ...

Hebrews 3:14 For we have come to share in Christ, if indeed we hold our original confidence firm to the end.

2 Timothy 1:7 For God gave us a spirit not of fear but of power and love and self-control.

Colossians 2:1-23 For I want you to know how great a struggle I have for you and for those at Laodicea and for all who have not seen me face to face, that their hearts may be encouraged, being knit together in love, to reach all the riches of full assurance of understanding and the knowledge of God's mystery, which is Christ, in whom are hidden all the treasures of wisdom and knowledge. I say this in order that no one may delude you with plausible arguments. For though I am absent in body, yet I am with you in spirit, rejoicing to see your good order and the firmness of your faith in Christ. ...

Colossians 1:5 Because of the hope laid up for you in heaven. Of this you have heard before in the word of the truth, the gospel,

Romans 12:1 I appeal to you therefore, brothers, by the mercies of God, to present your bodies as a living sacrifice, holy and acceptable to God, which is your spiritual worship.

Romans 5:2-5Through him we have also obtained access by faith into this grace in which we stand, and we rejoice in hope of the glory of God. More than that, we rejoice in our sufferings, knowing that suffering produces endurance, and endurance produces character, and character produces hope, and hope does not put us to shame, because God's love has been poured into our hearts through the Holy Spirit who has been given to us.

John 16:33 I have said these things to you, that in me you may have peace. In the world you will have tribulation. But take heart; I have overcome the world."

John 3:16-21 "For God so loved the world, that he gave his only Son, that whoever believes in him should not perish but have eternal life. For God did not send his Son into the world to condemn the world, but in order that the world might be saved through him. Whoever believes in him is not condemned, but whoever does not believe is condemned already, because he has not believed in the name of the only Son of God. And this is the judgment: the light has come into the world, and people loved the darkness rather than the light because their works were evil. For everyone who does wicked things hates the light and does not come to the light, lest his works should be exposed. ...

Matthew 28:19 Go therefore and make disciples of all nations, baptizing them in the name of the Father and of the Son and of the Holy Spirit,

Matthew 5:44 But I say to you, Love your enemies and pray for those who persecute you,

Jeremiah 29:11 For I know the plans I have for you, declares the Lord, plans for welfare and not for evil, to give you a future and a hope.

Isaiah 1:1-31 The vision of Isaiah the son of Amoz, which he saw concerning Judah and Jerusalem in the days of Uzziah, Jotham, Ahaz, and Hezekiah, kings of Judah. Hear, O heavens, and give ear, O earth; for the Lord has spoken: "Children have I reared and brought up, but they have rebelled against me. The ox knows its owner, and the donkey its master's crib, but Israel does not know, my people do not understand." Ah, sinful nation, a people laden with iniquity, offspring of evildoers, children who deal corruptly! They have forsaken the Lord, they have despised the Holy One of Israel, they are utterly estranged. Why will you still be struck down? Why will you continue to rebel? The whole head is sick, and the whole heart faint. ...

Psalm 119:41 Let your steadfast love come to me, O Lord, your salvation according to your promise;

Psalm 51:1-2 To the choirmaster. A Psalm of David, when Nathan the prophet went to him, after he had gone in to Bathsheba. Have mercy on me, O God, according to your steadfast love; according to your abundant mercy blot out my transgressions. Wash me thoroughly from my iniquity, and cleanse me from my sin!

1 Chronicles 16:11 Seek the Lord and his strength; seek his presence continually!

2 Peter 3:9The Lord is not slow to fulfill his promise as some count slowness, but is patient toward you, not wishing that any should perish, but that all should reach repentance.

Galatians 3:26 For in Christ Jesus you are all sons of God, through faith.

1 Corinthians 15:52In a moment, in the twinkling of an eye, at the last trumpet. For the trumpet will sound, and the dead will be raised imperishable, and we shall be changed.

1 Corinthians 15:51 Behold! I tell you a mystery. We shall not all sleep, but we shall all be changed,

1 Corinthians 15:20 But in fact Christ has been raised from the dead, the first fruits of those who have fallen asleep.

1 Corinthians 6:19-20 Or do you not know that your body is a temple of the Holy Spirit within you, whom you have from God? You are not your own, for you were bought with a price. So glorify God in your body.

John 6:44 No one can come to me unless the Father who sent me draws him. And I will raise him up on the last day.

Matthew 7:12 “So whatever you wish that others would do to you, do also to them, for this is the Law and the Prophets.

Isaiah 59:2 But your iniquities have made a separation between you and your God, and your sins have hidden his face from you so that he does not hear.

Proverbs 3:5 Trust in the Lord with all your heart, and do not lean on your own understanding.

Psalm 57:7 My heart is steadfast, O God, my heart is steadfast! I will sing and make melody!

Revelation 3:20 Behold, I stand at the door and knock. If anyone hears my voice and opens the door, I will come in to him and eat with him, and he with me.

Revelation 1:1-20 The revelation of Jesus Christ, which God gave him to show to his servants the things that must soon take place. He made it known by sending his angel to his servant John, who bore witness to the word of God and to the testimony of Jesus Christ, even to all that he saw. Blessed is the one who reads aloud the words of this prophecy, and blessed are those who hear, and who keep what is written in it, for the time is near. John to the seven churches that are in Asia: Grace to you and peace from him who is and who was and who is to come, and from the seven spirits who are before his throne, and from Jesus Christ the faithful witness, the firstborn of the dead, and the ruler of kings on earth. To him who loves us and has freed us from our sins by his blood ...

1 John 4:19 We love because he first loved us.

1 John 4:8 Anyone who does not love does not know God, because God is love.

1 John 4:7 Beloved, let us love one another, for love is from God, and whoever loves has been born of God and knows God.

1 John 2:24 Let what you heard from the beginning abide in you. If what you heard from the beginning abides in you, then you too will abide in the Son and in the Father.

2 Peter 3:14 Therefore, beloved, since you are waiting for these, be diligent to be found by him without spot or blemish, and at peace.

1 Peter 5:10 And after you have suffered a little while, the God of all grace, who has called you to his eternal glory in Christ, will himself restore, confirm, strengthen, and establish you.

1 Peter 5:8-9 prowls around like a roaring lion, seeking someone to devour. Resist him, firm in your faith, knowing that the same kinds of suffering are being experienced by your brotherhood throughout the world.

James 2:26 For as the body apart from the spirit is dead, so also faith apart from works is dead.

Hebrews 12:1 Therefore, since we are surrounded by so great a cloud of witnesses, let us also lay aside every weight, and sin which clings so closely, and let us run with endurance the race that is set before us,

Hebrews 7:1-28 For this Melchizedek, king of Salem, priest of the Most High God, met Abraham returning from the slaughter of the kings and blessed him, and to him Abraham apportioned a tenth part of everything. He is first, by translation of his name, king of righteousness, and then he is also king of Salem, that is, king of peace. He is without father or mother or genealogy, having neither beginning of days nor end of life, but resembling the Son of God he continues a priest forever. See how great this man was to whom Abraham the patriarch gave a tenth of the spoils! And those descendants of Levi who receive the priestly office have a commandment in the law to take tithes from the people, that is, from their brothers, though these also are descended from Abraham. ...

2 Thessalonians 3:3 But the Lord is faithful. He will establish you and guard you against the evil one.

Acts 2:1 When the day of Pentecost arrived, they were all together in one place.

2 Chronicles 27:6 So Jotham became mighty, because he ordered his ways before the Lord his God.

Jude 1:7 Just as Sodom and Gomorrah and the surrounding cities, which likewise indulged in sexual immorality and pursued unnatural desire, serve as an example by undergoing a punishment of eternal fire.

1 John 2:20 But you have been anointed by the Holy One, and you all have knowledge.

1 Peter 5:6-7 Humble yourselves, therefore, under the mighty hand of God so that at the proper time he may exalt you, casting all your anxieties on him, because he cares for you.

James 1:2 Count it all joy, my brothers, when you meet trials of various kinds,

Hebrews 6:19 We have this as a sure and steadfast anchor of the soul, a hope that enters into the inner place behind the curtain,

Titus 3:5 He saved us, not because of works done by us in righteousness, but according to his own mercy, by the washing of regeneration and renewal of the Holy Spirit,

2 Timothy 2:15 Do your best to present yourself to God as one approved, a worker who has no need to be ashamed, rightly handling the word of truth.

1 Timothy 1:15 The saying is trustworthy and deserving of full acceptance, that Christ Jesus came into the world to save sinners, of whom I am the foremost.

PERSEVERANCE

Bible verses about perseverance

One word in Christianity that is not emphasized enough is perseverance. It is not those who at one time in their life prayed a prayer to accept Christ and later fall away who will enter God's Kingdom. A true child of God will persevere in faith in Christ and it is these people who will enter Heaven.

Scripture makes it clear that God lives inside of believers and He will work in your life until the end.

God will use trials that happen in your life for good. While doing God's will He will hold you up. Fix your eyes on Christ, not the world nor your problems.

You will not get through your walk of faith without prayer. Jesus gave us parables to teach us that we should not stop knocking on God's door.

We should not lose hope. We've all been there praying for weeks, months, and even years for something.

Perseverance in prayer shows seriousness. I've seen God answer prayers in a matter of days and for some He answered a couple of years down the road. God is doing a good work in us that we don't see. Are you willing to wrestle with God?

God answers at the best time and in the best way. We should not only persist in prayers during trials, but also when everything is going good. We should be prayer warriors praying for our families, ways to advance God's kingdom, guidance, daily strength, help, giving thanks, etc. Remain steadfast! Perseverance builds character and a closer relationship with the Lord.

THINGS WE NEED TO PERSEVERE IN.

- Faith in Christ
- Witnessing to others
- Prayer
- The Christian lifestyle
- Suffering

QUOTES

- "Prayer is the acid test of the inner man's strength. A strong spirit is capable of praying much and praying with all perseverance until the answer comes. A weak one grows weary and fainthearted in the maintenance of praying." Watchman Nee

- "Our motto must continue to be perseverance. And ultimately I trust the Almighty will crown our efforts with success." William Wilberforce
- "By perseverance the snail reached the ark." Charles Spurgeon

WHAT DOES THE BIBLE SAY?

2 Peter 1:5-7 For this very reason, make every effort to add to your faith goodness; and to goodness, knowledge; and to knowledge, self-control; and to self-control, perseverance; and to perseverance, godliness; and to godliness, mutual affection; and to mutual affection, love.

1 Timothy 6:12 Fight the good fight of faith, lay hold on eternal life, whereunto thou art also called, and hast professed a good profession before many witnesses.

2 Timothy 4:7-8 I have fought the good fight, I have finished the race, and I have remained faithful. And now the prize awaits me—the crown of righteousness, which the Lord, the righteous Judge, will give me on the day of his return. And the prize is not just for me but for all who eagerly look forward to his appearing.

FOCUS ON CHRIST AND THE ETERNAL PRIZE.

Hebrews 12:1-3 Since we are surrounded by so many examples of faith, we must get rid of everything that slows us down, especially sin that distracts us. We must run the race that lies ahead of us and never give up. We must focus on Jesus, the source and goal of our faith. He saw the joy ahead of him, so he endured death on the cross and ignored the disgrace it brought him. Now he holds the honored position—the one next to God the Father on the heavenly throne. Think about Jesus, who endured opposition from sinners, so that you don't become tired and give up.

Philippians 3:14 I press on to reach the end of the race and receive the heavenly prize for which God, through Christ Jesus, is calling us.

Driving through the storm!James 1:2-5 My brothers and sisters, when you have many kinds of troubles, you should be full of joy, because you know that these troubles test your faith, and this will give you patience. Let your patience show itself perfectly in what you do. Then you will be perfect and complete and will have everything you need. But if any of you needs wisdom, you should ask God for it. He is generous to everyone and will give you wisdom without criticizing you.

Romans 5:2-4 Because of our faith, Christ has brought us into this place of undeserved privilege where we now stand, and we confidently and joyfully look forward to sharing God's glory. We can rejoice, too, when we run into problems and trials, for we know that they help us develop endurance. And endurance develops strength of character, and character strengthens our confident hope of salvation.

1 Peter 5:10-11 In his kindness God called you to share in his eternal glory by means of Christ Jesus. So after you have suffered a little while, he will restore, support, and strengthen you, and he will place you on a firm foundation. All power to him forever! Amen.

James 1:12 God blesses those who patiently endure testing and temptation. Afterward they will receive the crown of life that God has promised to those who love him.

WE MUST CONTINUE IN PRAYER.

Luke 11:5-9 Then, teaching them more about prayer, he used this story: "Suppose you went to a friend's house at midnight, wanting to borrow three loaves of bread. You say to him, A friend of mine has just arrived for a visit, and I have nothing for him to eat. And suppose he calls out from his bedroom, 'Don't bother me. The door is locked for the night, and my family and I are all in bed. I can't help you.' But I tell you this—though he won't do it for friendship's sake, if you keep knocking long enough, he will get up and give you whatever you need because of your shameless persistence. "And so I tell you, keep on asking, and you will receive what you ask for. Keep on seeking, and you will find. Keep on knocking, and the door will be opened to you.

Romans 12:12 Be happy in your confidence, be patient in trouble, and pray continually. Acts 1:14 They all joined together constantly in prayer, along with the women and Mary the mother of Jesus, and with his brothers.

Do not grow weary!

Galatians 6:9-10 So let's not get tired of doing what is good. At just the right time we will reap a harvest of blessing if we don't give up. Therefore, whenever we have the opportunity, we should do good to everyone–especially to those in the family of faith.

1Thessalonians 3:13 But ye, brethren, be not weary in well doing.

BE STRONG IN THE LORD.

2 Chronicles 15:7 Be ye strong, therefore, and do not let your hands be weak, for your work shall be rewarded.

Joshua 1:9 9 See that I command thee to be strong and of a good courage; be not afraid, neither be thou dismayed; for I, the LORD thy God, am with thee wherever thou goest.

1 Corinthians 16:13 Watch ye, stand fast in the faith, be brave, be strong.

Psalm 23:4 Even though I walk through the valley of the shadow of death, I will fear no evil, for you are with me; your rod and your staff, they comfort me.

Job 17:9 The righteous keep moving forward, and those with clean hands become stronger and stronger.

1 Corinthians 13:7 Love never gives up, never loses faith, is always hopeful, and endures through every circumstance. Lamentations 3:25-26 The Lord is good to those who depend on him, to those who search for him. So it is good to wait quietly for salvation from the Lord.

James 4:10 Humble yourselves before the Lord, and he will lift you up.

2 Thessalonians 1:2-4 Grace and peace to you from God the Father and the Lord Jesus Christ. We ought always to thank God for you, brothers and sisters, and rightly so, because your faith is growing more and more, and the love all of you have for one another is increasing. Therefore, among God's churches we boast about your perseverance and faith in all the persecutions and trials you are enduring.

Revelation 2:2-3 I know your deeds, your hard work and your perseverance. I know that you cannot tolerate wicked people, that you have tested those who claim to be apostles but are not, and have found them false. You have persevered and have endured hardships for my name, and have not grown weary.

25. James 5:11 As you know, we count as blessed those who have persevered. You have heard of Job's perseverance and have seen what the Lord finally brought about. The Lord is full of compassion and mercy.

2 Corinthians 4:17 For our light affliction, which is but for a moment, worketh for us a far more exceeding and eternal weight of glory.

BIBLE SCRIPTURES ON FOCUS.

Proverbs 16:3 Commit your work to the Lord, and your plans will be established.

Romans 12:2 Do not be conformed to this world, but be transformed by the renewal of your mind, that by testing you may discern what is the will of God, what is good and acceptable and perfect.

Matthew 24:13"But the one who endures to the end will be saved."

Philippians 4:13I can do all things through him who strengthens me.

Romans 12:21 Do not be overcome by evil, but overcome evil with good.

1 Corinthians 2:14 The natural person does not accept the things of the Spirit of God, for they are folly to him, and he is not able to understand them because they are spiritually discerned.

1 Peter 2:24 He himself bore our sins in his body on the tree, that we might die to sin and live to righteousness. By his wounds you have been healed.

1 Peter 5:8 Be sober-minded; be watchful. Your adversary the devil prowls around like a roaring lion, seeking someone to devour.

2 Timothy 1:7 For God gave us a spirit not of fear but of power and love and self-control.

2 Timothy 2:4 No soldier gets entangled in civilian pursuits, since his aim is to please the one who enlisted him.

Colossians 3:2 Set your minds on things that are above, not on things that are on earth.

Ephesians 4:29 Let no corrupting talk come out of your mouths, but only such as is good for building up, as fits the occasion, that it may give grace to those who hear.

Hebrews 12:2 Looking to Jesus, the founder and perfecter of our faith, who for the joy that was set before him endured the cross, despising the shame, and is seated at the right hand of the throne of God.

Isaiah 26:3 You keep him in perfect peace whose mind is stayed on you, because he trusts in you.

Isaiah 41:10"Fear not, for I am with you; be not dismayed, for I am your God; I will strengthen you, I will help you, I will uphold you with my righteous right hand."

Isaiah 42:5 Thus says God, the Lord, who created the heavens and stretched them out, who spread out the earth and what comes from it, who gives breath to the people on it and spirit to those who walk in it.

Isaiah 50:7 But the Lord God helps me; therefore I have not been disgraced; therefore I have set my face like a flint, and I know that I shall not be put to shame.

Joshua 1:8"This Book of the Law shall not depart from your mouth, but you shall meditate on it day and night, so that you may be careful to do according to all that is written in it. For then you will make your way prosperous, and then you will have good success."

Luke 21:36"But stay awake at all times, praying that you may have strength to escape all these things that are going to take place, and to stand before the Son of Man."

Matthew 24:22"And if those days had not been cut short, no human being would be saved. But for the sake of the elect those days will be cut short."

Matthew 24:42"Therefore, stay awake, for you do not know on what day your Lord is coming."

Matthew 24:44"Therefore you also must be ready, for the Son of Man is coming at an hour you do not expect."

Matthew 6:24"No one can serve two masters, for either he will hate the one and love the other, or he will be devoted to the one and despise the other. You cannot serve God and money."

Matthew 6:33"But seek first the kingdom of God and his righteousness, and all these things will be added to you."

Nehemiah 6:3 And I sent messengers to them, saying, "I am doing a great work and I cannot come down. Why should the work stop while I leave it and come down to you?"

Philippians 4:8 Finally, brothers, whatever is true, whatever is honorable, whatever is just, whatever is pure, whatever is lovely, whatever is commendable, if there is any excellence, if there is anything worthy of praise, think about these things.

Proverbs 4:25 Let your eyes look directly forward, and your gaze be straight before you.

Psalm 119:15I will meditate on your precepts and fix my eyes on your ways.

Romans 13:11 Besides this you know the time, that the hour has come for you to wake from sleep. For salvation is nearer to us now than when we first believed.

Romans 15:4 For whatever was written in former days was written for our instruction, that through endurance and through the encouragement of the Scriptures we might have hope

BIBLE SCRIPTURES ON PATIENCE

Romans 12:12 Rejoice in hope, be patient in tribulation, be constant in prayer.

Galatians 6:9 And let us not grow weary of doing good, for in due season we will reap, if we do not give up.

Romans 8:25 But if we hope for what we do not see, we wait for it with patience.

Ephesians 4:2 With all humility and gentleness, with patience, bearing with one another in love.

Philippians 4:6 Do not be anxious about anything, but in everything by prayer and supplication with thanksgiving let your requests be made known to God.

1 Corinthians 13:4 Love is patient and kind; love does not envy or boast; it is not arrogant.

1 Thessalonians 5:14 And we urge you, brothers, admonish the idle, encourage the fainthearted, help the weak, be patient with them all.

1 Timothy 1:16 But I received mercy for this reason, that in me, as the foremost, Jesus Christ might display his perfect patience as an example to those who were to believe in him for eternal life.

2 Peter 3:9 The Lord is not slow to fulfill his promise as some count slowness, but is patient toward you, not wishing that any should perish, but that all should reach repentance.

2 Timothy 2:24 And the Lord's servant must not be quarrelsome but kind to everyone, able to teach, patiently enduring evil.

2 Timothy 4:2 Preach the word; be ready in season and out of season; reprove, rebuke, and exhort, with complete patience and teaching.

Colossians 1:11 May you be strengthened with all power, according to his glorious might, for all endurance and patience with joy.

Colossians 3:12 Put on then, as God's chosen ones, holy and beloved, compassionate hearts, kindness, humility, meekness, and patience.

Ecclesiastes 7:8 Better is the end of a thing than its beginning, and the patient in spirit is better than the proud in spirit.

Ecclesiastes 7:9 Be not quick in your spirit to become angry, for anger lodges in the bosom of fools.

Galatians 5:22
But the fruit of the Spirit is love, joy, peace, patience, kindness, goodness, faithfulness.

Hebrews 10:36 For you have need of endurance, so that when you have done the will of God you may receive what is promised.

Hebrews 12:1 Therefore, since we are surrounded by so great a cloud of witnesses, let us also lay aside every weight, and sin which clings so closely, and let us run with endurance the race that is set before us.

Hebrews 6:12 So that you may not be sluggish, but imitators of those who through faith and patience inherit the promises.

Hebrews 6:15 And thus Abraham, having patiently waited, obtained the promise.

Isaiah 40:31 But they who wait for the Lord shall renew their strength; they shall mount up with wings like eagles; they shall run and not be weary; they shall walk and not faint.

James 1:19 Know this, my beloved brothers: let every person be quick to hear, slow to speak, slow to anger.

James 1:3 For you know that the testing of your faith produces steadfastness.

James 1:4 And let steadfastness have its full effect, that you may be perfect and complete, lacking in nothing.

James 5:7 Be patient, therefore, brothers, until the coming of the Lord. See how the farmer waits for the precious fruit of the earth, being patient about it, until it receives the early and the late rains.

James 5:8 You also, be patient. Establish your hearts, for the coming of the Lord is at hand.

Jeremiah 29:11 For I know the plans I have for you, declares the Lord, plans for welfare and not for evil, to give you a future and a hope.

Joel 2:13 And rend your hearts and not your garments. Return to the Lord your God, for he is gracious and merciful, slow to anger, and abounding in steadfast love; and he relents over disaster.

Luke 21:19 By your endurance you will gain your lives.

Luke 8:15 As for that in the good soil, they are those who, hearing the word, hold it fast in an honest and good heart, and bear fruit with patience.

Proverbs 14:29 Whoever is slow to anger has great understanding, but he who has a hasty temper exalts folly.

Proverbs 15:18A hot-tempered man stirs up strife, but he who is slow to anger quiets contention.

Proverbs 16:32 Whoever is slow to anger is better than the mighty, and he who rules his spirit than he who takes a city.

Proverbs 25:15 With patience a ruler may be persuaded, and a soft tongue will break a bone.

Psalm 37:7 Be still before the Lord and wait patiently for him; fret not yourself over the one who prospers in his way, over the man who carries out evil devices!

Psalm 40:1 To the choirmaster. A Psalm of David. I waited patiently for the Lord; he inclined to me and heard my cry.

Romans 15:5 May the God of endurance and encouragement grant you to live in such harmony with one another, in accord with Christ Jesus.

Romans 2:4 Or do you presume on the riches of his kindness and forbearance and patience, not knowing that God's kindness is meant to lead you to repentance?

Romans 5:3 More than that, we rejoice in our sufferings, knowing that suffering produces endurance.

Romans 5:4 And endurance produces character, and character produces hope.

MOST POWERFUL BIBLE SCRIPTURES ON FAITH

The word "faith" occurs 514 times in the Bible. The infographic below identifies the points of each occurrence. This graphic presentation starts with Genesis in the top left and concludes with Revelation in the bottom right.

Matthew 21:22"And whatever you ask in prayer, you will receive, if you have faith."

Romans 10:17 So faith comes from hearing, and hearing through the word of Christ.

Hebrews 11:6 And without faith it is impossible to please him, for whoever would draw near to God must believe that he exists and that he rewards those who seek him.

Hebrews 11:1 Now faith is the assurance of things hoped for, the conviction of things not seen.

James 2:19 You believe that God is one; you do well. Even the demons believe—and shudder!

1 Corinthians 13:13 So now faith, hope, and love abide, these three; but the greatest of these is love.

1 Corinthians 13:2
And if I have prophetic powers, and understand all mysteries and all knowledge, and if I have all faith, so as to remove mountains, but have not love, I am nothing.

1 Corinthians 16:13 Be watchful, stand firm in the faith, act like men, be strong.

1 Corinthians 2:5 That your faith might not rest in the wisdom of men but in the power of God.

1 John 5:1 Everyone who believes that Jesus is the Christ has been born of God, and everyone who loves the Father loves whoever has been born of him.

1 John 5:4 For everyone who has been born of God overcomes the world. And this is the victory that has overcome the world—our faith.

1 Peter 3:15 But in your hearts honor Christ the Lord as holy, always being prepared to make a defense to anyone who asks you for a reason for the hope that is in you; yet do it with gentleness and respect.

1 Timothy 6:12 Fight the good fight of the faith. Take hold of the eternal life to which you were called and about which you made the good confession in the presence of many witnesses.

2 Corinthians 5:7 For we walk by faith, not by sight.

2 Timothy 4:7I have fought the good fight, I have finished the race, I have kept the faith.

Ephesians 2:8 For by grace you have been saved through faith. And this is not your own doing; it is the gift of God.

Ephesians 6:16 In all circumstances take up the shield of faith, with which you can extinguish all the flaming darts of the evil one.

Galatians 2:16 Yet we know that a person is not justified by works of the law but through faith in Jesus Christ, so we also have believed in Christ Jesus, in order to be justified by faith in Christ and not by works of the law, because by works of the law no one will be justified.

Galatians 2:20I have been crucified with Christ. It is no longer I who live, but Christ who lives in me. And the life I now live in the flesh I live by faith in the Son of God, who loved me and gave himself for me.

Habakkuk 2:4
Behold, his soul is puffed up; it is not upright within him, but the righteous shall live by his faith.

Hebrews 11:7 By faith Noah, being warned by God concerning events as yet unseen, in reverent fear constructed an ark for the saving of his household. By this he condemned the world and became an heir of the righteousness that comes by faith.

Hebrews 12:2 Looking to Jesus, the founder and perfecter of our faith, who for the joy that was set before him endured the cross, despising the shame, and is seated at the right hand of the throne of God.

James 1:3 For you know that the testing of your faith produces steadfastness.

James 1:6 But let him ask in faith, with no doubting, for the one who doubts is like a wave of the sea that is driven and tossed by the wind.

James 2:17 So also faith by itself, if it does not have works, is dead.

James 2:24 You see that a person is justified by works and not by faith alone.

John 1:1 In the beginning was the Word, and the Word was with God, and the Word was God.

John 3:16 For God so loved the world, that he gave his only Son, that whoever believes in him should not perish but have eternal life.

John 3:36"Whoever believes in the Son has eternal life; whoever does not obey the Son shall not see life, but the wrath of God remains on him."

John 6:35 Jesus said to them, "I am the bread of life; whoever comes to me shall not hunger, and whoever believes in me shall never thirst."

John 8:24"I told you that you would die in your sins, for unless you believe that I am he you will die in your sins."

Luke 1:37"For nothing will be impossible with God."

Luke 17:5 The apostles said to the Lord, "Increase our faith!"

Luke 18:27 But he said, "What is impossible with men is possible with God."

Mark 10:52 And Jesus said to him, "Go your way; your faith has made you well." And immediately he recovered his sight and followed him on the way.

Mark 11:22 And Jesus answered them, "Have faith in God."

Mark 11:24"Therefore I tell you, whatever you ask in prayer, believe that you have received it, and it will be yours."

Mark 16:16"Whoever believes and is baptized will be saved, but whoever does not believe will be condemned."

Mark 9:23 And Jesus said to him, "If you can! All things are possible for one who believes."

Matthew 15:28 Then Jesus answered her, "O woman, great is your faith! Be it done for you as you desire." And her daughter was healed instantly.

Matthew 17:20 He said to them, "Because of your little faith. For truly, I say to you, if you have faith like a grain of mustard seed, you will say to this mountain, 'Move from here to there,' and it will move, and nothing will be impossible for you."

Matthew 21:21 And Jesus answered them, "Truly, I say to you, if you have faith and do not doubt, you will not only do what has been done to the fig tree, but even if you say to this mountain, 'Be taken up and thrown into the sea,' it will happen."

Matthew 6:24"No one can serve two masters, for either he will hate the one and love the other, or he will be devoted to the one and despise the other. You cannot serve God and money."

Philippians 4:13I can do all things through him who strengthens me.

Proverbs 3:5 Trust in the Lord with all your heart, and do not lean on your own understanding.

Psalm 46:10"Be still, and know that I am God. I will be exalted among the nations, I will be exalted in the earth!"

Romans 1:17 For in it the righteousness of God is revealed from faith for faith, as it is written, "The righteous shall live by faith."

Romans 10:9 Because, if you confess with your mouth that Jesus is Lord and believe in your heart that God raised him from the dead, you will be saved.

Romans 12:3
For by the grace given to me I say to everyone among you not to think of himself more highly than he ought to think, but to think with sober judgment, each according to the measure of faith that God has assigned.

Romans 15:13 May the God of hope fill you with all joy and peace in believing, so that by the power of the Holy Spirit you may abound in hope

BIBLE SCRIPTURES ON TRUST IN THE LORD

The word "trust" occurs 153 times in the Bible. The info graphic below identifies the points of each occurrence. This graphic presentation starts with Genesis in the top left and concludes with Revelation in the bottom right.

Proverbs 3:5
Trust in the Lord with all your heart, and do not lean on your own understanding.

Jeremiah 29:11"For I know the plans I have for you," declares the Lord, "plans for welfare and not for evil, to give you a future and a hope."

1 John 4:18 There is no fear in love, but perfect love casts out fear. For fear has to do with punishment, and whoever fears has not been perfected in love.

Psalm 37:5 Commit your way to the Lord; trust in him, and he will act.

Mark 11:24"Therefore I tell you, whatever you ask in prayer, believe that you have received it, and it will be yours."

1 John 4:16 So we have come to know and to believe the love that God has for us. God is love, and whoever abides in love abides in God, and God abides in him.

1 John 5:14 And this is the confidence that we have toward him, that if we ask anything according to his will he hears us.

2 Corinthians 5:7 For we walk by faith, not by sight.

Hebrews 11:6 And without faith it is impossible to please him, for whoever would draw near to God must believe that he exists and that he rewards those who seek him.

Hebrews 13:6 So we can confidently say, "The Lord is my helper; I will not fear; what can man do to me?"

Isaiah 41:10 Fear not, for I am with you; be not dismayed, for I am your God; I will strengthen you, I will help you, I will uphold you with my righteous right hand.

Isaiah 43:2 When you pass through the waters, I will be with you; and through the rivers, they shall not overwhelm you; when you walk through fire you shall not be burned, and the flame shall not consume you.

James 1:6 But let him ask in faith, with no doubting, for the one who doubts is like a wave of the sea that is driven and tossed by the wind.

Jeremiah 17:5
Thus says the Lord: "Cursed is the man who trusts in man and makes flesh his strength, whose heart turns away from the Lord."

Jeremiah 17:7“Blessed is the man who trusts in the Lord, whose trust is the Lord.”

Job 13:15 Though he slay me, I will hope in him; yet I will argue my ways to his face.

John 10:10“The thief comes only to steal and kill and destroy. I came that they may have life and have it abundantly.”

John 14:1“Let not your hearts be troubled. Believe in God; believe also in me.”

John 14:27“Peace I leave with you; my peace I give to you. Not as the world gives do I give to you. Let not your hearts be troubled, neither let them be afraid.”

John 3:16 For God so loved the world, that he gave his only Son, that whoever believes in him should not perish but have eternal life.

Joshua 1:9“Have I not commanded you? Be strong and courageous. Do not be frightened, and do not be dismayed, for the Lord your God is with you wherever you go.”

Luke 16:10“One who is faithful in a very little is also faithful in much, and one who is dishonest in a very little is also dishonest in much.”

Mark 5:36 But overhearing what they said, Jesus said to the ruler of the synagogue, “Do not fear, only believe.”

Mark 9:24 Immediately the father of the child cried out and said, “I believe; help my unbelief!”

Nahum 1:7 The Lord is good, a stronghold in the day of trouble; he knows those who take refuge in him.

Philippians 4:19 And my God will supply every need of yours according to his riches in glory in Christ Jesus.

Proverbs 16:20 Whoever gives thought to the word will discover good, and blessed is he who trusts in the Lord.

Proverbs 16:3 Commit your work to the Lord, and your plans will be established.

Proverbs 28:25A greedy man stirs up strife, but the one who trusts in the Lord will be enriched.

Proverbs 29:25 The fear of man lays a snare, but whoever trusts in the Lord is safe.

Proverbs 3:6 In all your ways acknowledge him, and he will make straight your paths.

Proverbs 30:5 Every word of God proves true; he is a shield to those who take refuge in him.

Psalm 112:7 He is not afraid of bad news; his heart is firm, trusting in the Lord.

Psalm 118:8 It is better to take refuge in the Lord than to trust in man.

Psalm 13:5 But I have trusted in your steadfast love; my heart shall rejoice in your salvation.

Psalm 143:8 Let me hear in the morning of your steadfast love, for in you I trust. Make me know the way I should go, for to you I lift up my soul.

Psalm 146:3 Put not your trust in princes, in a son of man, in whom there is no salvation.

Psalm 20:7 Some trust in chariots and some in horses, but we trust in the name of the Lord our God.

Psalm 27:13I believe that I shall look upon the goodness of the Lord in the land of the living!

Psalm 28:7 The Lord is my strength and my shield; in him my heart trusts, and I am helped; my heart exults, and with my song I give thanks to him.

Psalm 31:14 But I trust in you, O Lord; I say, "You are my God."

Psalm 32:10 Many are the sorrows of the wicked, but steadfast love surrounds the one who trusts in the Lord.

Psalm 37:3 Trust in the Lord, and do good; dwell in the land and befriend faithfulness.

Psalm 4:5 Offer right sacrifices, and put your trust in the Lord.

Psalm 40:4 Blessed is the man who makes the Lord his trust, who does not turn to the proud, to those who go astray after a lie!

Psalm 46:10"Be still, and know that I am God. I will be exalted among the nations, I will be exalted in the earth!"

Psalm 55:22 Cast your burden on the Lord, and he will sustain you; he will never permit the righteous to be moved.

Psalm 56:3 When I am afraid, I put my trust in you.

Psalm 56:4 In God, whose word I praise, in God I trust; I shall not be afraid. What can flesh do to m?

Psalm 60:12 With God we shall do valiantly; it is he who will tread down our foes.

Psalm 62:8 Trust in him at all times, O people; pour out your heart before him; God is a refuge for us. Selah

Psalm 9:10 And those who know your name put their trust in you, for you, O Lord, have not forsaken those who seek you.

Romans 10:9 Because, if you confess with your mouth that Jesus is Lord and believe in your heart that God raised him from the dead, you will be saved.

Romans 12:2 Do not be conformed to this world, but be transformed by the renewal of your mind, that by testing you may discern what is the will of God, what is good and acceptable and perfect.

Romans 15:13 May the God of hope fill you with all joy and peace in believing, so that by the power of the Holy Spirit you may abound in hope.

Romans 8:28 And we know that for those who love God all things work together for good, for those who are called according to his purpose.

Titus 3:5 He saved us, not because of works done by us in righteousness, but according to his own mercy, by the washing of regeneration and renewal of the Holy Spirit same yesterday and today and forever.

Isaiah 26:3 You keep him in perfect peace whose mind is stayed on you, because he trusts in you.

POWERFUL BIBLE SCRIPTURES ON FORGIVENESS

Here are the 68 most powerful Bible scriptures on forgiveness. We have organized these incredible Bible verses by book.

The word "forgive" appears 127 times throughout the books of the Bible. The graph below shows the locations of each appearance. It begins with Genesis in the top left and concludes with Revelation in the bottom right.

POPULAR BIBLE VERSES ABOUT FORGIVENESS

Ephesians 4:32 Be kind to one another, tenderhearted, forgiving one another, as God in Christ forgave you.

Mark 11:25 And whenever you stand praying, forgive, if you have anything against anyone, so that your Father also who is in heaven may forgive you your trespasses.

1 John 1:9 If we confess our sins, he is faithful and just to forgive us our sins and to cleanse us from all unrighteousness.

Matthew 18:21 Then Peter came up and said to him, "Lord, how often will my brother sin against me, and I forgive him? As many as seven times?" Jesus said to him, "I do not say to you seven times, but seventy times seven."

Luke 6:37 Judge not, and you will not be judged; condemn not, and you will not be condemned; forgive, and you will be forgiven.

MORE BELOVED SCRIPTURES ON THE THEME OF FORGIVENESS

1 Corinthians 10:13 No temptation has overtaken you that is not common to man. God is faithful, and he will not let you be tempted beyond your ability, but with the temptation he will also provide the way of escape, that you may be able to endure it.

1 Corinthians 13:4-7 Love is patient and kind; love does not envy or boast; it is not arrogant or rude. It does not insist on its own way; it is not irritable or resentful; it does not rejoice at wrongdoing, but rejoices with the truth. Love bears all things, believes all things, hopes all things, endures all things.

1 John 1:8 If we say we have no sin, we deceive ourselves, and the truth is not in us.

1 John 2:1 My little children, I am writing these things to you so that you may not sin. But if anyone does sin, we have an advocate with the Father, Jesus Christ the righteous.

1 John 2:12I am writing to you, little children, because your sins are forgiven for his name's sake.

1 John 2:2 He is the propitiation for our sins, and not for ours only but also for the sins of the whole world.

1 Peter 4:8 Above all, keep loving one another earnestly, since love covers a multitude of sins.

2 Chronicles 7:14 If my people who are called by my name humble themselves, and pray and seek my face and turn from their wicked ways, then I will hear from heaven and will forgive their sin and heal their land.

2 Corinthians 5:17 Therefore, if anyone is in Christ, he is a new creation. The old has passed away; behold, the new has come.

2 Peter 3:9 The Lord is not slow to fulfill his promise as some count slowness, but is patient toward you, not wishing that any should perish, but that all should reach repentance.

Acts 10:43 To him all the prophets bear witness that everyone who believes in him receives forgiveness of sins through his name.

Acts 13:38 Let it be known to you therefore, brothers, that through this man forgiveness of sins is proclaimed to you.

Acts 17:30 The times of ignorance God overlooked, but now he commands all people everywhere to repent.

Acts 2:38 And Peter said to them, "Repent and be baptized every one of you in the name of Jesus Christ for the forgiveness of your sins, and you will receive the gift of the Holy Spirit."

Acts 3:19 Repent therefore, and turn again, that your sins may be blotted out.

Colossians 1:14 In whom we have redemption, the forgiveness of sins.

Colossians 3:13 Bearing with one another and, if one has a complaint against another, forgiving each other; as the Lord has forgiven you, so you also must forgive.

Colossians 3:23 Whatever you do, work heartily, as for the Lord and not for men.

Daniel 9:9 To the Lord our God belong mercy and forgiveness, for we have rebelled against him.

Ephesians 1:7 In him we have redemption through his blood, the forgiveness of our trespasses, according to the riches of his grace.

Galatians 6:1 Brothers, if anyone is caught in any transgression, you who are spiritual should restore him in a spirit of gentleness. Keep watch on yourself, lest you too be tempted.

Genesis 50:17'Say to Joseph, Please forgive the transgression of your brothers and their sin, because they did evil to you.' And now, please forgive the transgression of the servants of the God of your father." Joseph wept when they spoke to him.

Hebrews 10:17 Then he adds, "I will remember their sins and their lawless deeds no more."

Hebrews 4:16 Let us then with confidence draw near to the throne of grace, that we may receive mercy and find grace to help in time of need.

Hebrews 8:12 For I will be merciful toward their iniquities, and I will remember their sins no more.

Hebrews 9:22 Indeed, under the law almost everything is purified with blood, and without the shedding of blood there is no forgiveness of sins.

Isaiah 1:18 Come now, let us reason together, says the Lord: though your sins are like scarlet, they shall be as white as snow; though they are red like crimson, they shall become like wool.

Isaiah 43:25I, I am he who blots out your transgressions for my own sake, and I will not remember your sins.

Isaiah 53:5 But he was wounded for our transgressions; he was crushed for our iniquities; upon him was the chastisement that brought us peace, and with his stripes we are healed.

Isaiah 55:7 Let the wicked forsake his way, and the unrighteous man his thoughts; let him return to the Lord, that he may have compassion on him, and to our God, for he will abundantly pardon.

James 4:7ubmit yourselves therefore to God. Resist the devil, and he will flee from you.

James 5:15 And the prayer of faith will save the one who is sick, and the Lord will raise him up. And if he has committed sins, he will be forgiven.

James 5:16 Therefore, confess your sins to one another and pray for one another, that you may be healed. The prayer of a righteous person has great power as it is working.

Jeremiah 31:34 And no longer shall each one teach his neighbor and each his brother, saying, 'Know the Lord,' for they shall all know me, from the least of them to the greatest, declares the Lord. For I will forgive their iniquity, and I will remember their sin no more.

Joel 2:13 And rend your hearts and not your garments. Return to the Lord your God, for he is gracious and merciful, slow to anger, and abounding in steadfast love; and he relents over disaster.

John 3:16 For God so loved the world, that he gave his only Son, that whoever believes in him should not perish but have eternal life.

John 8:7 And as they continued to ask him, he stood up and said to them, "Let him who is without sin among you be the first to throw a stone at her."

Leviticus 19:18 You shall not take vengeance or bear a grudge against the sons of your own people, but you shall love your neighbor as yourself: I am the Lord.

Leviticus 19:28 You shall not make any cuts on your body for the dead or tattoo yourselves: I am the Lord.

Luke 12:33 Sell your possessions, and give to the needy. Provide yourselves with moneybags that do not grow old, with a treasure in the heavens that does not fail, where no thief approaches and no moth destroys.

Luke 23:34 And Jesus said, "Father, forgive them, for they know not what they do." And they cast lots to divide his garments.

Luke 6:27 But I say to you who hear, Love your enemies, do good to those who hate you.

Mark 3:28-29"Truly, I say to you, all sins will be forgiven the children of man, and whatever blasphemies they utter, but whoever blasphemes against the Holy Spirit never has forgiveness, but is guilty of an eternal sin."

Matthew 11:28 Come to me, all who labor and are heavy laden, and I will give you rest.

Matthew 18:15 If your brother sins against you, go and tell him his fault, between you and him alone. If he listens to you, you have gained your brother.

Matthew 18:35 So also my heavenly Father will do to every one of you, if you do not forgive your brother from your heart.

Matthew 26:28 For this is my blood of the covenant, which is poured out for many for the forgiveness of sins.

Matthew 5:44 But I say to you, Love your enemies and pray for those who persecute you,

Matthew 5:7 Blessed are the merciful, for they shall receive mercy.

Matthew 6:12 And forgive us our debts, as we also have forgiven our debtors.

Matthew 6:14 For if you forgive others their trespasses, your heavenly Father will also forgive you.

Matthew 6:15 But if you do not forgive others their trespasses, neither will your Father forgive your trespasses.

Micah 7:18 Who is a God like you, pardoning iniquity and passing over transgression for the remnant of his inheritance? He does not retain his anger forever, because he delights in steadfast love.

Proverbs 10:12 Hatred stirs up strife, but love covers all offenses.

Proverbs 17:9 Whoever covers an offense seeks love, but he who repeats a matter separates close friends.

Proverbs 28:13 Whoever conceals his transgressions will not prosper, but he who confesses and forsakes them will obtain mercy.

Psalm 103:12 As far as the east is from the west, so far does he remove our transgressions from us.

Psalm 32:5I acknowledged my sin to you, and I did not cover my iniquity; I said, "I will confess my transgressions to the Lord," and you forgave the iniquity of my sin.

Psalm 86:5 For you, O Lord, are good and forgiving, abounding in steadfast love to all who call upon you.

Romans 12:20 To the contrary, "if your enemy is hungry, feed him; if he is thirsty, give him something to drink; for by so doing you will heap burning coals on his head."

Romans 3:23 For all have sinned and fall short of the glory of God.

Romans 5:8 But God shows his love for us in that while we were still sinners, Christ died for us.

Romans 6:23 For the wages of sin is death, but the free gift of God is eternal life in Christ Jesus our Lord.

Romans 8:1 There is therefore now no condemnation for those who are in Christ Jesus.

<u>TENACITY: THE WILLINGNESS TO PERSIST</u>

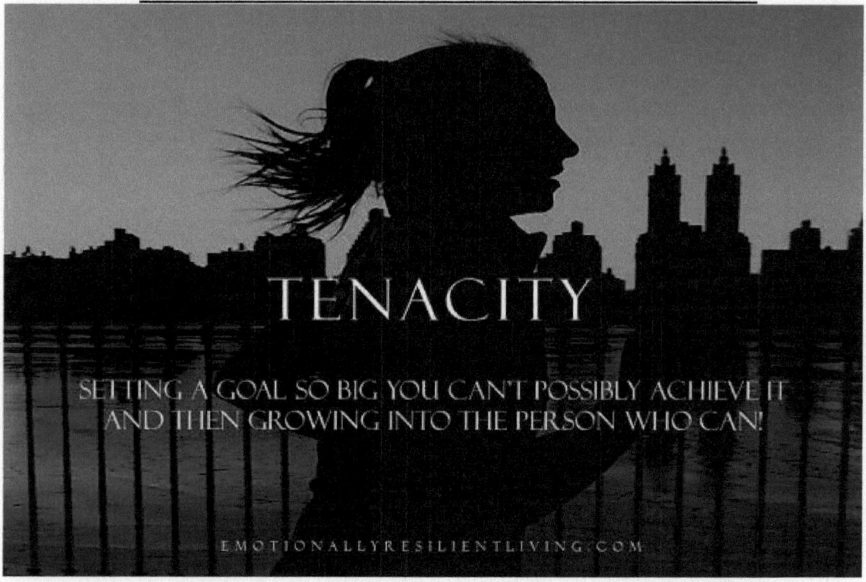

Determined, persistent, tenacious, perseverance, stick-to-itiveness – whatever label you apply it's about hanging on through the challenges and bumps in the road to eventually achieve your goals.

Now, not to put too fine a point on it, but while each of these words represents the same fundamental ideology, I will admit to having a strong affinity for the concept of ***tenacity***.

It seems to me that tenacity is about so much more than not giving up on a goal. It's about pushing through barriers of doubt and self-limiting beliefs, and most importantly, having the willingness to persist despite all odds.

Persistence is doing something again and again until it works. Tenacity is using new data to make new decisions to find new pathways to find new ways to achieve goals when the old ways didn't work.~Seth Godin

Standard goal setting advice subscribes to the importance of being realistic, and there's a lot to be said for that approach when it comes to many areas of life.

But here's what you need to keep in mind. Through the years we each develop beliefs about our own limitations, what we are capable of having, being or doing in life.

In reality, you are capable of so much more than you realize!

Where you'll achieve the most meaningful growth is by challenging yourself to go a little beyond what is comfortable. When you see that you can do more, you'll be more inclined to risk reaching just a little bit further.

So what does tenacity look like?

Tenacious people are planners, and while they are disciplined and apply an unwavering approach to going after their desired outcome, they clearly understand the difference between not giving up and stubbornly doing the same thing over and over hoping for a different outcome.

I've seen stubbornness referred to as tenacity's ugly twin and I think that sums it up pretty well.

Those who are tenacious focus on growth with purpose so they continually evaluate their actions and progress; remain flexible and adjust their methods and leave nothing to chance.

They are masters at self motivation, avid learners and have no qualms about asking for help when the need arises.

The most difficult thing is the decision to act the rest is merely tenacity. The fears are paper tigers. You can do anything you decide to do. You can act to change and control your life; and the procedure, the process is its own reward.~Robyn Davidson

But I want to add a cautionary note here.

The initial excitement of setting a big, audacious goal is pretty heady stuff so it's understandable that you might want to share your exciting plans and enthusiasm.

If you're fortunate enough to have forward thinking people in your support network, that's great.

However many find themselves faced with naysayers warning of the foolishness of dreaming big dreams.

I experienced this firsthand in a very big way when, while working as administrative assistant to a hotel general manager in Hawaii, I decided I wanted to go into sales.

The list of all the reasons why that was not only an impossible but reckless idea was daunting, but the worst part was that none of my friends or family supported my dream, not one.

Everyone told me repeatedly that I was being foolish by risking a "steady and reliable" job pursuing a pipedream.

Thankfully, my co-workers and professional peers provided the support network I needed and while I will admit it was not an easy journey, I managed to weather the storm and grew well beyond my goal to earn a senior level position in international sales and eventually went on to spearhead the marketing and educational programs for a billion-dollar tourist attraction.

My point in sharing this with you is that above all those who are tenacious aren't just hard workers, they have a clear and compelling vision that serves as an anchor no matter what challenges come their way.

CLOSING THOUGHTS

Let's be honest, this notion of going for BIG unrealistic dreams is not for everyone and that really is okay. But for those who dare, it's also important to bear in mind that such adventures are rarely undertaken in a vacuum.

We begin by going for a goal that is just a little out of reach, something that requires that we must obtain new knowledge and learn new skills in order to achieve, and then continue to build on each successive accomplishment.

This is an approach that can be mastered by any determined individual and it is surprisingly easy to know if and when you're ready to make the leap because it inevitably boils down to just how badly you want to achieve something, to be uncommon.

How about you? Do you tenaciously pursue your dreams and goals? Do you ever set BIG goals that require you to reach beyond your comfort zone?

Will you be the passenger or the driver in your life journey?

BIBLIOGRAPHY

1. Jump up to:[a b c d] *"Letter 5140 – Wallace, A. R. to Darwin, C. R., 2 July 1866". Darwin Correspondence Project. Retrieved 12 January 2010.*
   "Letter 5145 – Darwin, C. R. to Wallace, A. R., 5 July (1866)". Darwin Correspondence Project. Retrieved 12 January 2010.
   ^ "Herbert Spencer in his *Principles of Biology* of 1864, vol. 1, p. 444, wrote: 'This survival of the fittest, which I have here sought to express in mechanical terms, is that which Mr. Darwin has called "natural selection", or the preservation of favoured races in the struggle for life.'" *Maurice E. Stucke, Better Competition Advocacy, retrieved 29 August 2007*, citing HERBERT SPENCER, THE PRINCIPLES OF BIOLOGY 444 (Univ. Press of the Pac. 2002.)
2. ^ Jump up to:[a b] "This preservation, during the battle for life, of varieties which possess any advantage in structure, constitution, or instinct, I have called Natural Selection; and Mr. Herbert Spencer has well expressed the same idea by the Survival of the Fittest. The term "natural selection" is in some respects a bad one, as it seems to imply conscious choice; but this will be disregarded after a little familiarity." *Darwin, Charles (1868), The Variation of Animals and Plants under Domestication, **1** (1st ed.), London: John Murray, p. 6, retrieved 10 August 2015*
3. ^ Jump up to:[a b] *Freeman, R. B. (1977), "On the Origin of Species", The Works of Charles Darwin: An Annotated Bibliographical Handlist (2nd ed.), Cannon House, Folkestone, Kent, England: Wm Dawson & Sons Ltd*
4. ^ Jump up to:[a b] "This preservation of favourable variations, and the destruction of injurious variations, I call Natural Selection, or the Survival of the Fittest." – *Darwin, Charles (1869), On the Origin of Species by Means of Natural Selection, or the Preservation of Favoured Races in the Struggle for Life (5th ed.), London: John Murray, pp. 91–92, retrieved 22 February 2009*
5. ^ Jump up to:[a b c] "Stephen Jay Gould, Darwin's Untimely Burial", 1976; from Philosophy of Biology:An Anthology, Alex Rosenberg, Robert Arp ed., John Wiley & Sons, May 2009, pp. 99–102.
6. **^** "Evolutionary biologists customarily employ the metaphor 'survival of the fittest,' which has a precise meaning in the context of mathematical population genetics, as a shorthand expression when describing evolutionary processes." *Chew, Matthew K.; Laubichler, Manfred D. (4 July 2003), "PERCEPTIONS OF SCIENCE: Natural*

Enemies – Metaphor or Misconception?", Science, ***301*** *(5629): 52–53, doi:10.1126/science.1085274, PMID 12846231*

7. **^** *Johnson, Curtis (2014). Darwin's Dice: The Idea of Chance in the Thought of Charles Darwin. Oxford University Press. p. 155. ISBN 9780199361434.*
8. **^** Vol. 1, p. 444
9. ^ Jump up to:[a b c d] *"Letter 5140 – Wallace, A. R. to Darwin, C. R., 2 July 1866". Darwin Correspondence Project. Retrieved 12 January 2010.*
   "Letter 5145 – Darwin, C. R. to Wallace, A. R., 5 July (1866)". Darwin Correspondence Project. Retrieved 12 January 2010.
   ^ "Herbert Spencer in his *Principles of Biology* of 1864, vol. 1, p. 444, wrote: 'This survival of the fittest, which I have here sought to express in mechanical terms, is that which Mr. Darwin has called "natural selection", or the preservation of favoured races in the struggle for life.'" *Maurice E. Stucke, Better Competition Advocacy, retrieved 29 August 2007*, citing HERBERT SPENCER, THE PRINCIPLES OF BIOLOGY 444 (Univ. Press of the Pac. 2002.)
10. ^ Jump up to:[a b] "This preservation, during the battle for life, of varieties which possess any advantage in structure, constitution, or instinct, I have called Natural Selection; and Mr. Herbert Spencer has well expressed the same idea by the Survival of the Fittest. The term "natural selection" is in some respects a bad one, as it seems to imply conscious choice; but this will be disregarded after a little familiarity." *Darwin, Charles (1868), The Variation of Animals and Plants under Domestication,* ***1*** *(1st ed.), London: John Murray, p. 6, retrieved 10 August 2015*
11. ^ Jump up to:[a b] *Freeman, R. B. (1977), "On the Origin of Species", The Works of Charles Darwin: An Annotated Bibliographical Handlist (2nd ed.), Cannon House, Folkestone, Kent, England: Wm Dawson & Sons Ltd*
12. ^ Jump up to:[a b] "This preservation of favourable variations, and the destruction of injurious variations, I call Natural Selection, or the Survival of the Fittest." – *Darwin, Charles (1869), On the Origin of Species by Means of Natural Selection, or the Preservation of Favoured Races in the Struggle for Life (5th ed.), London: John Murray, pp. 91–92, retrieved 22 February 2009*
13. ^ Jump up to:[a b c] "Stephen Jay Gould, Darwin's Untimely Burial", 1976; from Philosophy of Biology:An Anthology, Alex

Rosenberg, Robert Arp ed., John Wiley & Sons, May 2009, pp. 99–102.
14. ^ "Evolutionary biologists customarily employ the metaphor 'survival of the fittest,' which has a precise meaning in the context of mathematical population genetics, as a shorthand expression when describing evolutionary processes." *Chew, Matthew K.; Laubichler, Manfred D. (4 July 2003), "PERCEPTIONS OF SCIENCE: Natural Enemies – Metaphor or Misconception?", Science,* ***301*** *(5629): 52–53, doi:10.1126/science.1085274, PMID 12846231*
15. ^ *Johnson, Curtis (2014). Darwin's Dice: The Idea of Chance in the Thought of Charles Darwin. Oxford University Press. p. 155. ISBN 9780199361434.*
16. ^ Vol. 1, p. 444

MIX
Papier aus verantwortungsvollen Quellen
Paper from responsible sources
FSC® C105338

Printed by Books on Demand GmbH, Norderstedt / Germany